THINGS THAT FLOAT

Annabel Thomas

Designed by Steve Page

Illustrated by Peter Bull

Contents

All about ships and boats

This book tells you about lots of different ships and boats. You can find out how they float, what makes them go and what they do. It also tells you about some unusual ships and boats around the world, as well as those that have broken records for speed or size.

Cargo ships, like this one, carry goods from one port to another. See pages 14-15.

This is a power boat. You can find out how it works on pages 8-9.

A hovercraft skims over the water on a cushion of air. Pages 16-17 tell you how.

You can find out about different kinds of sailing boats on pages 10-11.

Small submarines, like this one, are called submersibles. See pages 18-19.

Boats and their uses

Some large ships, called liners, are built for holiday cruises. They are like floating hotels.

Huge naval ships have a runway on the deck where planes can take off and land.

Some ferries have ramps so cars, trucks and buses can drive on and off them.

Lifeboats are specially built and equipped to save people from drowning at sea.

Floating and sinking

A big steel or wooden boat is very heavy. When it floats it pushes some of the water aside. The water round the boat pushes back. This push (force) of the water holds up the boat if the boat is not too heavy for its size.

A heavy boat needs to have high sides so lots of water can push against it.

Wood is light for its size and floats easily. ▼

Although metal is heavy, a steel ship can float. ▶

Hull

Because it is hollow, the metal shell or hull of a boat weighs less than a solid amount of metal of the same size. Both push aside the same amount of water, but the force of the water can support the weight of the hull because it is hollow.

Clay boat

Try this experiment.

Make the sides thin. ▶

Take two pieces of clay of the same weight. Roll one into a ball and press out the other to make a cup-shaped boat.

Basin

Put them into some water. The ball sinks. Although both weigh the same, the "boat" floats as there is more water pushing it up.

Boat power

The earliest way of moving a boat along was by a person's muscles, using paddles or oars.

For thousands of years, sails were used to catch the wind and push the boat forward.

Now most large ships have diesel engines, but some use steam turbine engines.

Many modern submarines use nuclear power to turn steam turbines.

Steamboats

The first engines were invented at the end of the 18th century and were powered by steam. Here you can see an early steamboat engine. The steam is used to turn paddle-wheels either side of the boat. As they turn, they push against the water. This makes the boat move forward.

What is steam?

Steam is very hot and can burn you.

Steam is a gas which is made when water boils. You can see the powerful way it spouts from a boiling kettle. It is this power which works a steam-engine.

Steamships

Savannah

In 1819, the first sailing ship with an engine crossed the Atlantic. An American ship, *Savannah*, only used its engines for 8 hours of the 21 day journey.

Sirius

The first steamship to make a crossing without using its sails was the British ship, *Sirius*, built in 1837. It took 18 days.

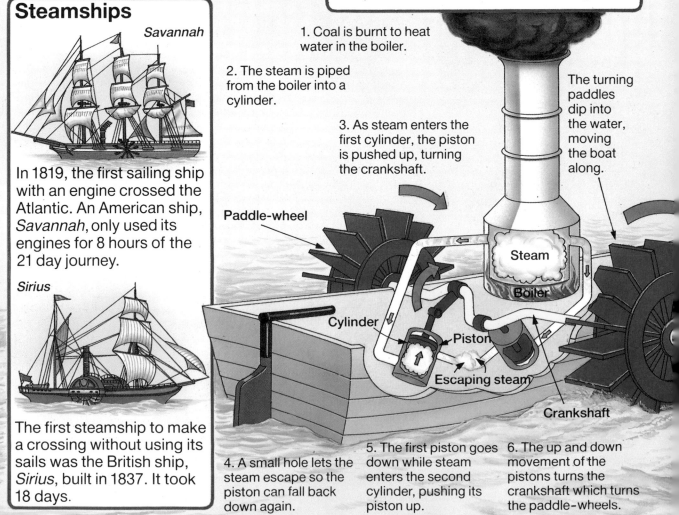

1. Coal is burnt to heat water in the boiler.

2. The steam is piped from the boiler into a cylinder.

3. As steam enters the first cylinder, the piston is pushed up, turning the crankshaft.

The turning paddles dip into the water, moving the boat along.

Paddle-wheel

Steam

Boiler

Cylinder

Piston

Escaping steam

Crankshaft

4. A small hole lets the steam escape so the piston can fall back down again.

5. The first piston goes down while steam enters the second cylinder, pushing its piston up.

6. The up and down movement of the pistons turns the crankshaft which turns the paddle-wheels.

Screw propulsion

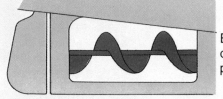

Early type of screw propeller.

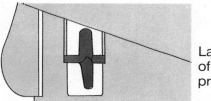

Later type of screw propeller.

In the 1840s screw propellers were invented and fitted to the back of boats. Although they were smaller than paddle-wheels they made the boat go faster. Like paddle-wheels, they were powered by steam engines and were turned by the spinning motion of the crankshaft.

During the trial run of a boat, the end of a long screw propeller broke off. It worked better so propellers were made shorter.

How propellers work

The *Great Britain* was the first iron ship to use a screw propeller rather than paddle-wheels.

Propellers work by going through the water like a corkscrew goes through a cork. As the blades of the propeller turn, the water is forced backwards. This thrusts the boat forwards.

Steam turbines

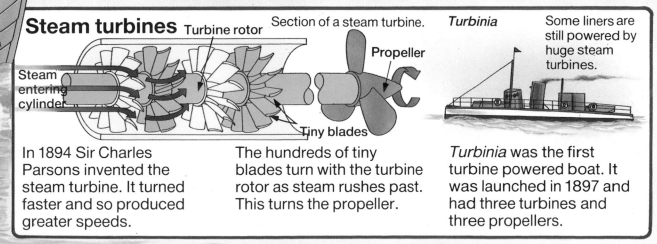

Turbine rotor

Section of a steam turbine.

Propeller

Steam entering cylinder

Tiny blades

Turbinia

Some liners are still powered by huge steam turbines.

In 1894 Sir Charles Parsons invented the steam turbine. It turned faster and so produced greater speeds.

The hundreds of tiny blades turn with the turbine rotor as steam rushes past. This turns the propeller.

Turbinia was the first turbine powered boat. It was launched in 1897 and had three turbines and three propellers.

Liners

Liners are very big ships. In the past people had to travel by sea if they wanted to go to another country. Nowadays airplanes take much of the passenger traffic and liners are used mostly for holiday cruises.

The biggest passenger liner in use is the *Queen Elizabeth II,* or *QE2*. It is like a small city, with shops, restaurants, movies and even a hospital.

These are kennels where passengers can keep their pets.

The *QE2*

In the ship's shopping center, passengers can buy things, such as clothes, food and flowers.

This is the ship's theater.

There are four swimming pools.

Tennis court

The *QE2* has two propellers, each with six blades.

The ship's health club is here.

This is the ship's laundry.

This is the ship's control room – a huge computer works out its speed and direction.

This is the turbine room, where all the ship's diesel powered turbine engines are.

Measuring speed

At sea, speed is measured in knots. The name comes from the time when a sailor would throw the end of a knotted line into sea. As the ship moved forward, the line unravelled and the knots, which were equally spaced, were counted over a period of time.

The average travelling speed of the *QE2* is 28½ knots.

Nowadays a knot is a speed of one nautical mile per hour. A nautical (sea) mile is different from a land mile. It is 1852 meters, so a knot is 1.85 kilometers an hour (1.15 miles an hour).

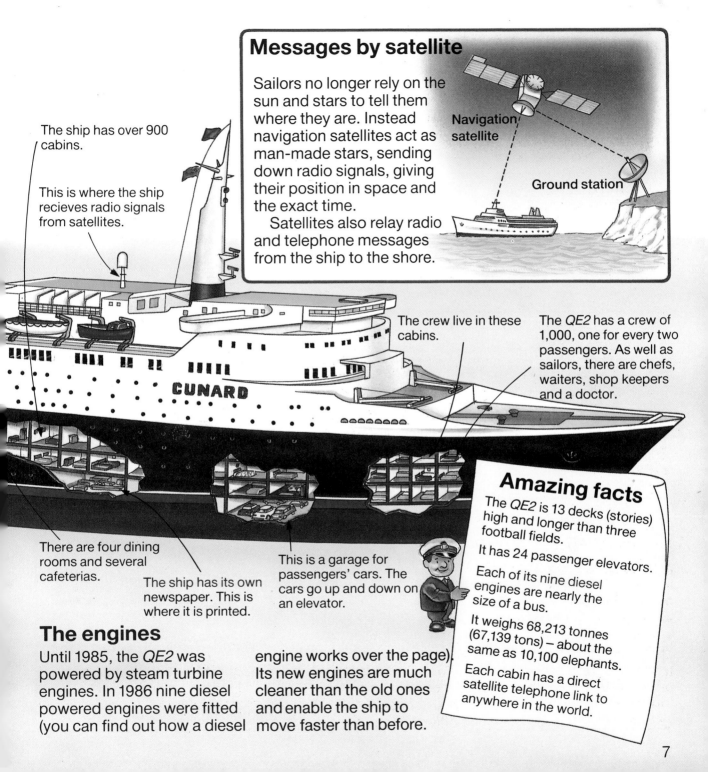

The ship has over 900 cabins.

This is where the ship recieves radio signals from satellites.

Messages by satellite

Sailors no longer rely on the sun and stars to tell them where they are. Instead navigation satellites act as man-made stars, sending down radio signals, giving their position in space and the exact time.

Satellites also relay radio and telephone messages from the ship to the shore.

Navigation satellite

Ground station

The crew live in these cabins.

The QE2 has a crew of 1,000, one for every two passengers. As well as sailors, there are chefs, waiters, shop keepers and a doctor.

CUNARD

There are four dining rooms and several cafeterias.

The ship has its own newspaper. This is where it is printed.

This is a garage for passengers' cars. The cars go up and down on an elevator.

Amazing facts

The QE2 is 13 decks (stories) high and longer than three football fields.

It has 24 passenger elevators.

Each of its nine diesel engines are nearly the size of a bus.

It weighs 68,213 tonnes (67,139 tons) – about the same as 10,100 elephants.

Each cabin has a direct satellite telephone link to anywhere in the world.

The engines

Until 1985, the QE2 was powered by steam turbine engines. In 1986 nine diesel powered engines were fitted (you can find out how a diesel engine works over the page). Its new engines are much cleaner than the old ones and enable the ship to move faster than before.

Boats and their engines

Most small boats have a gasoline engine fixed at the back. They are called outboard motors because the engine can be lifted off the boat. Larger boats have powerful diesel engines, housed inside the hull. They are called inboard motors.

Each engine is protected by a case.

This boat is called a power boat. It has two diesel engines and two propellers.

Panel showing boat's speed and how much fuel there is.

Steering wheel

Deck

Long, narrow hull for racing.

The engine

A diesel engine, like a gasoline engine, is called an internal combustion engine. This means the fuel is burnt inside the engine. You can see how on the next page.

Pilot boats

Pilot boats take sea pilots to large ships. The pilot then guides the ship through dangerous and unfamiliar water.

Motor cruisers

Motor cruisers are often big boats, with living quarters on board. They are used for cruising holidays.

Tenders

A tender, such as this one, is kept aboard large ocean liners for carrying passengers from the ship to the shore.

The diesel engine

The diesel engine was invented by a German, Dr Rudolph Diesel, in 1897. It uses a special type of fuel called diesel oil.

A boat engine can have between two and twelve cylinders. The more cylinders an engine has, the greater its power.

This engine has four cylinders. Each cylinder shows one of four things that happen to turn the propeller.

2. The valve closes and the piston moves up, squashing the air. This makes the air very hot.

3. The injector squirts fuel into the hot air. The mixture explodes, pushing the piston down.

1. Air enters the cylinder through the inlet valve and the piston moves down.

Injector

Inlet valve

Cylinder

Exhaust valve

Piston

Crankshaft

Modern propellers have twisted blades.

4. The exhaust valve opens and the piston pushes the waste gases out.

The propeller

The up and down movement of the pistons turns the crankshaft which turns the propeller. Its blades push the water backwards and the boat is driven forwards.

Power boat racing

Power boats are designed specially for racing. The fastest have jet engines, the kind of engine an airplane uses.

A famous racing event is the Bahamas Powerboat Grand-Prix.

Record breaker

It took 3 days, 8 hours and 31 minutes to cross a distance of over 5,000 km (3,000 miles).

In 1986, a British power boat, *Virgin Atlantic Challenger*, crossed the Atlantic in record time. It won the Blue Riband, previously awarded to liners for the fastest Atlantic crossing.

Sailing boats

For thousands of years boats with sails have relied on the power of the wind to push them along. The sails "catch" the wind and the force of the wind pushing against the sails moves the boat forward.

An arrangement of sails is called a rig. You can find out about the development of rigs on the opposite page.

The first sails

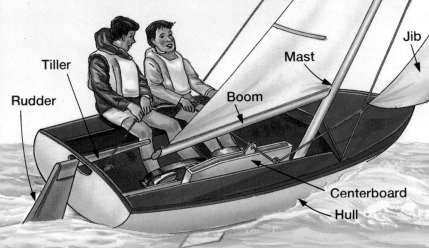

wind

wind

About 5,000 years ago, the Egyptians used square sails. When the wind blew from behind the boat was pushed forward.

For thousands of years the Arabs used triangular (lateen) sails. They used ropes to curve the sails round to catch the wind.

Modern sails are triangular with a curved outside edge. This style is called Bermudan.

Small boats like this sailing dinghy only use a mainsail and a smaller sail, called the jib.

Steering the boat

The boat is steered by the tiller, which acts like a steering wheel. The tiller is joined to the rudder which changes the direction of the boat.

The mainsail is joined to the boom and mast.

The jib is an extra sail which catches the wind. It helps to steer the boat and turn it round.

Nowadays most sails are made of a light, waterproof material.

Mainsail

Tiller

Rudder

Mast

Boom

Jib

Centerboard

Hull

When the tiller is moved to the right, it moves the rudder and the boat turns left. When it is moved to the left, the boat goes right.

The centerboard keeps the boat going straight. It helps to stop the boat drifting sideways, when the wind pushes on the sails.

The catamaran

The trimaran

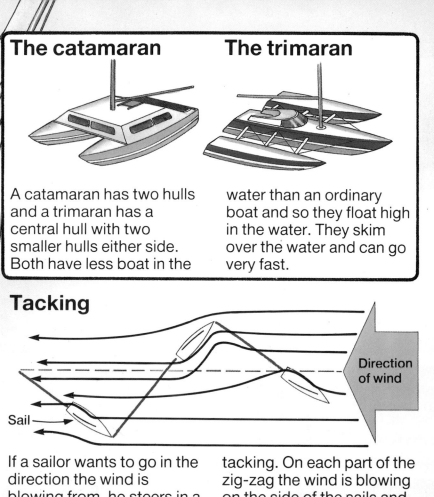

A catamaran has two hulls and a trimaran has a central hull with two smaller hulls either side. Both have less boat in the water than an ordinary boat and so they float high in the water. They skim over the water and can go very fast.

Tacking

Direction of wind

Sail

If a sailor wants to go in the direction the wind is blowing from, he steers in a zig-zag. This is called tacking. On each part of the zig-zag the wind is blowing on the side of the sails and pushes the boat forwards.

The America's Cup

The America's Cup is a yachting event which takes place every four years in the country of the last winner. In 1986-87 the races were held in Australia and were won by the American yacht *Stars and Stripes*.

America's Cup trophy

Story of rigs

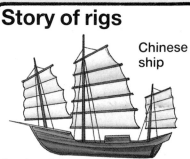

Chinese ship

In the 9th century, Chinese ships were built with several masts and sails made of bamboo matting. This design lasted for hundreds of years.

Three-masted ship

In the 15th century, three-masted ships were built in Europe. These ships were used for sea battles, exploring and trade.

Clipper

In the 1820s cargo ships called Clippers were made. They had many large sails, a long slim hull and could go very fast.

Muscle power

Long ago people used their hands like paddles to propel their boats along. Then they made wooden paddles which were bigger than their hands and worked better.

Later long oars were used, like the ones in the picture. These worked even more efficiently.

Rowing boats like this one are made from very light but strong material, such as fiberglass.

The cox shouts instructions to the crew and steers the boat.

Racing crews often practice several hours a day to make sure they work well together to make the boat go faster.

The cox

Pushing water aside

The way you move a paddle or oar through water is similar to the way you move your arms when you swim. As the paddle or oar pushes the water backwards, the boat moves forwards.

Rowlock

Each oar rests on a rowlock, so the oar works like a lever.

Ships with oars

Ancient Egyptian ship

Viking longship

Steering oar

Ancient Greek warship

As long as 5,000 years ago, the Egyptians rowed their ships along with oars when there was no wind or it was blowing the wrong way.

About 800 years ago, the people of Scandinavia, called Vikings, built long, narrow ships, called longships. They had up to 25 oars on each side.

The oars of an Ancient Greek warship were usually arranged on different levels so that the oarsmen did not get in each other's way.

Going faster

People soon discovered that the length of the oars was important. A single pull on a long oar pushes the boat further forward than a single pull on a short oar.

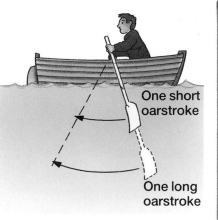

One short oarstroke

One long oarstroke

The boat's streamlined shape enables it to travel through the water at great speed.

Canoeing

Protective helmet

Double ended paddle

A type of canoe is still used in some countries, such as Alaska, for fishing and transport. Mostly, canoes are used for sporting events, such as the slalom. Competitors have to weave their canoe in and out of a row of poles in fast flowing water, without hitting them.

Unusual boats

Gondola

Gondolas are boats used on the canals of Venice in Italy. A gondolier stands at the back of the boat, propelling it along with one long oar.

Punt

Flat bottomed boats called punts are used for pleasure. A long pole is pushed against the river bed to propel the punt along.

Reed boat

Boats made of reeds are still used in some countries round the world, such as Peru in South America. People use long poles to push the boat along.

Cargo ships

Cargo ships carry goods, or cargo, from one port to another. The cargo can be anything from oranges to steel rods or coal to wheat.

A port has roads and railways to bring cargo to the ship. It also has huge warehouses where cargo can be stored before being loaded or unloaded on to or off a ship.

Containers

Warehouses

A ship's cargo is always checked to ensure nothing is smuggled in or out of a country.

Special crane for lifting containers.

Container ship

Crane with sling unloading boxes.

Grain being piped down a chute.

▼A container ship carries cargo in large boxes called containers. These are packed before being taken to the port.

Some ships carry cargo of all shapes and sizes. It has to be tightly packed so it cannot slide about at sea
◄ and be damaged.

Some ships carry bulk food, such as sugar or wheat. It is poured down chutes into the ship and sucked out again by pipes.

Roll on/roll off ships are built so loaded trucks and trains can drive straight on and off the ship, without delay.

Roll on/roll off ship

Railway

Road

Oil tankers

Very big ships that carry oil are called supertankers. They are too large for most docks so the oil is piped to storage tanks from special terminals outside the port.

Storage tank

One of the largest supertankers, *Globtik London*, is 378 meters (413 yards) long. The crew use bicycles to get around on the deck.

Changing direction

A tanker is very difficult to turn because of its size. The captain needs to know if there is anything the tanker is likely to hit, well before it is in sight. Radar signals, relayed to a computer are used to help the captain. You can see how on the right.

Ship being towed out of harbor by tugs.

Some tankers and other ships are so big they cannot be steered into ports. Small boats, called tugs, tow them in and out of the docks, and put them into position.

Using computers

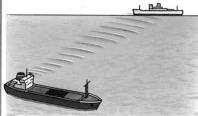

Tankers send out signals, called radar, to look for ships and rocks. Radar travels through the air until it "hits" something.

The signals bounce back to a radar screen. The tanker's computer works out how fast and in which direction the ship must go to avoid a collision.

Computer sails

Shin Aitoku Maru

The Japanese tanker, *Shin Aitoku Maru,* has special metal sails to push it along. A computer works out when the sails should be turned to catch the wind.

Things that skim

A hovercraft is a type of boat that skims above water or land supported by a cushion of air. It is sometimes called an air-cushion vehicle or ACV.

There are two other types of boats that skim above the water. They are the hydrofoil and the jetfoil. You can find out about them on the opposite page.

The hovercraft

The propellers spin pushing the hovercraft forwards.

Propellers

Floating on air

Yoghurt carton

Air

Styrofoam tray

To test how a hovercraft works, cut the bottom out of a yoghurt carton. Then cut a hole in the middle of a styrofoam tray big enough to put the yoghurt carton in. If you blow into the carton, the tray will move easily on a cushion of air.

Steering the hovercraft

The propellers pivot to steer the hovercraft and the rudders move to one side or the other when it changes direction.

A hovercraft keeps steady in rough water as the skirt can move up and down or bulge out when the waves push the cushion about.

Rudder

The hydrofoil

A hydrofoil is a boat that has underwater wings (called hydrofoils). The whole boat lifts out of the water as it gathers speed.

How a hydrofoil works

The top surface of the wings of a hydrofoil are smooth, so water quickly runs off them. The wings rise up, lifting the whole boat out of the water. There is then no water to push against the boat so it can go very fast.

Steam turbine engine

A fan under each propeller sucks in air to fill the base. Each fan and propeller are driven by a steam or gas turbine engine.

Fan

Skirt

A rubber skirt fitted round the base stops the cushion of air escaping.

Stopping at the port

A hovercraft comes out of the water. The engines are turned off and as no air goes into the skirt it sinks to rest on its base.

V-foils

Some hydrofoils have V shaped wings, called V-foils. They stick out of the water on both sides of the boat as it goes faster.

Submerged foils

Submerged foils stay under the water so the hydrofoil looks as if it has legs. They can change direction to suit different weather conditions.

The Jetfoil

A jetfoil is a type of hydrofoil. It is propelled forwards by two water jets. Gas turbines work the pumps which force the water through holes under great pressure to make the jets.

Direction of water

Submarines

Submarines travel under the sea. They are powered by diesel engines or nuclear powered turbines. A nuclear submarine, like the one in this picture, has a rounder hull than a submarine with a diesel engine. Nuclear submarines can work for years without needing to be refuelled and can stay under water for as long as two years without coming to the surface.

Radio antennae
The radio antennae pick up satellite messages.

The periscope
A periscope is a tube with a mirror at both ends. When it is raised a sailor in the submarine can see what is happening above the water while the rest of the submarine is below the water.

Hydroplanes

Conning tower
The submarine is steered from the conning tower.

Sonar detector
The sonar detector picks up sound waves (see opposite page).

Double hull

Ballast tanks

Control room

Sleeping quarters

Diving

Ballast tank

Before the submarine dives, shutters are opened so the sea floods into the ballast tanks and the submarine sinks.

Underwater

Once under the water, the level of water in the ballast tanks is adjusted so the submarine stays at a chosen depth.

Surfacing

To surface, air is forced into the ballast tanks under great pressure. This forces out the water and the submarine rises.

Propeller
The propeller make the submarine go forwards.

Hydroplanes
There are four hydroplanes, two at the front and two at the back. They help direct the submarine as it goes under the water.

Nuclear powered steam turbines

Submarines have a double hull. Between its two walls are ballast tanks – tanks that are filled with water to make the submarine sink.

Sonar

Sonar is a way of finding out where other ships and submarines are from the sounds they make. There are two kinds of sonar, Active and Passive.

Active sonar

The submarine sends out sound waves. When they hit something they "ping" and an echo bounces back to the submarine.

Passive sonar

Passive sonar picks up the smallest sound using electronic equipment. It makes no sound so the submarine's presence is secret.

The *Turtle*

The first submarine was built by an American in the late 18th century. It was shaped like an egg and had no engines.

Nautilus

In 1958, an American submarine, *Nautilus,* was the first vessel to reach the North Pole. It travelled there under the ice.

The Bathyscaphe

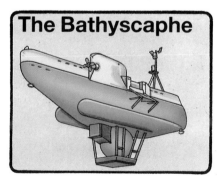

The Bathyscaphe is a submersible (small submarine) designed specially for very deep underwater research.

Lifeboats

Bad weather at sea often causes accidents and shipwrecks. Lifeboats are designed to go out in strong winds and rough seas, and their crews are trained to rescue people in danger of drowning.

Padded jacket
Waterproof padded jacket is worn for warmth. It is brightly colored to show up against the sea.

Waterproof trousers

Lifejacket
Lifejacket, filled with air, keeps a person afloat in the sea.

Bump cap
The bump cap and hood protects the head.

Inflatables

Inflatable lifeboats rescue people close to the shore.

Getting the right way up

1

2 Superstructure

The superstructure has watertight doors.

3

If a lifeboat capsizes, it can come back up again within a few seconds. This is called self-righting. A lifeboat does not sink

because air is trapped inside the superstructure (the top of the boat). The weight of the heavy engines

in the bottom of the boat then pulls the hull back into the water, so the boat is the right way up again.

Fishing boats

Fishing boats, called trawlers, have enormous nets, called trawls. This trawler is called a purse seiner. Its net circles the fish and is drawn in by a rope before being winched aboard.

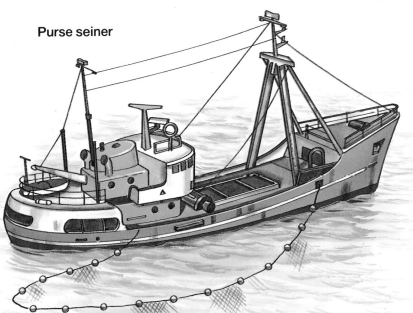

Purse seiner

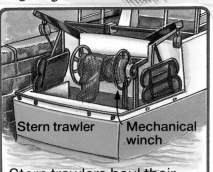

Stern trawler   Mechanical winch

Stern trawlers haul their nets in from the stern (back of the boat). They are hauled in by mechanical winch.

On board

Once on board, the fish is either packed with ice in boxes or put in huge freezers. Ships with freezers can stay at sea for a long time, without the fish going bad.

Some ships even have fish factories on board. You can find out about them on the right.

Factory ships

Factory ships have factories on board where the fish are cleaned and prepared for sale. Often smaller fishing boats off-load their haul onto factory ships at sea.

The fish are sent along big square pipes to huge, square trays. Here, a factory worker cleans and prepares the fish.

Some prepared fish are stored in barrels and then stacked on the ship's deck. Others are packed in trays and frozen.

Biggest and fastest

On this page you can find out about some of the world's biggest and fastest ships. Many of them are naval ships. Some are so big that they have runways on deck where aeroplanes can land and take off.

The biggest

Nimitz

Apart from some tankers, the biggest ships to have been built are three United States' Navy aircraft carriers called *Nimitz*, *Dwight D. Eisenhower* and *Carl Vinson*. Each weighs 92,869 tonnes (91,374 tons).

The flight deck is 1090 feet (333 meters) long and 252 feet (77 meters) wide.

Seawise Giant

The biggest oil tanker is called *Seawise Giant*. It is owned by Liberia, though it was built in Japan. It weighs 564,733 tonnes (555,697 tons) and is 1,504 feet (458 meters) long. It is so long that *Nimitz* is only two thirds its length.

The fastest

SS United States

Le Terrible

Bluebird

In 1967 an updated *Bluebird* overturned having reached a speed of 527 km/h (327 mph).

The fastest passenger liner was the *SS United States*. On its first voyage in 1952, it crossed the Atlantic at a speed of 35.59 knots (66 km/h, 41 mph).

A French destroyer (a light, fast warship), called *Le Terrible*, built in 1935 could travel as fast as 45.25 knots (83.9 km/h, 52.1 mph).

In 1956, *Bluebird*, a jet-propelled speedboat, reached a speed of over 360 km/h (223 mph) on an English lake. This is the world water speed record.

Sea-way code

At sea, there are rules that ships and boats have to obey, just as cars have to obey rules when on the road. Sailors learn to read signs and signals from other ships, lighthouses and buoys. Even though nowadays ships send messages by radio, sailors still learn all the old rules of the sea.

Lighthouse

Lighthouses

Lighthouses are built on rocky headlands. Their lights warn ships and boats to keep clear. They are also built at sea, to mark rocks and reefs.

When standing on the deck of a ship facing its bow (front), port is to the left and starboard to the right.

Lightships

A lightship is used in an area where it is impossible to build a lighthouse, for instance on a sandbank.

Lightship

Starboard

Port

Port

Starboard

Keep to the right

One of the international rules of the sea is to keep to the right. This rule stops ships and boats crashing into each other. These two ships are both going to the right, otherwise they would collide.

Both ships must also give one short blast on a siren.

Sea signals

Nowadays most ships send out radio signals to tell other ships of their whereabouts. They can also make strong blasts on a siren when it is foggy.

Buoys

Buoys mark areas that might be dangerous for boats, such as hidden rocks, channels or even wrecks of ships. Their positions are also marked on charts of the sea.

Buoy

Index

First published in 1987 by
Usborne Publishing Ltd,
20 Garrick Street,
London WC2E 9BJ,
England.

© 1987 Usborne Publishing Ltd.

The name Usborne and the device 🐝 are the Trade Marks of Usborne Publishing Ltd.

Printed in Belgium.

American edition 1987.